Copyright © 2019 by Ashton Stokes

All rights reserved. No part of this book may be reproduced or used in any manner without written permission of the copyright owner except for the use of quotations in a book review.

FIRST EDITION

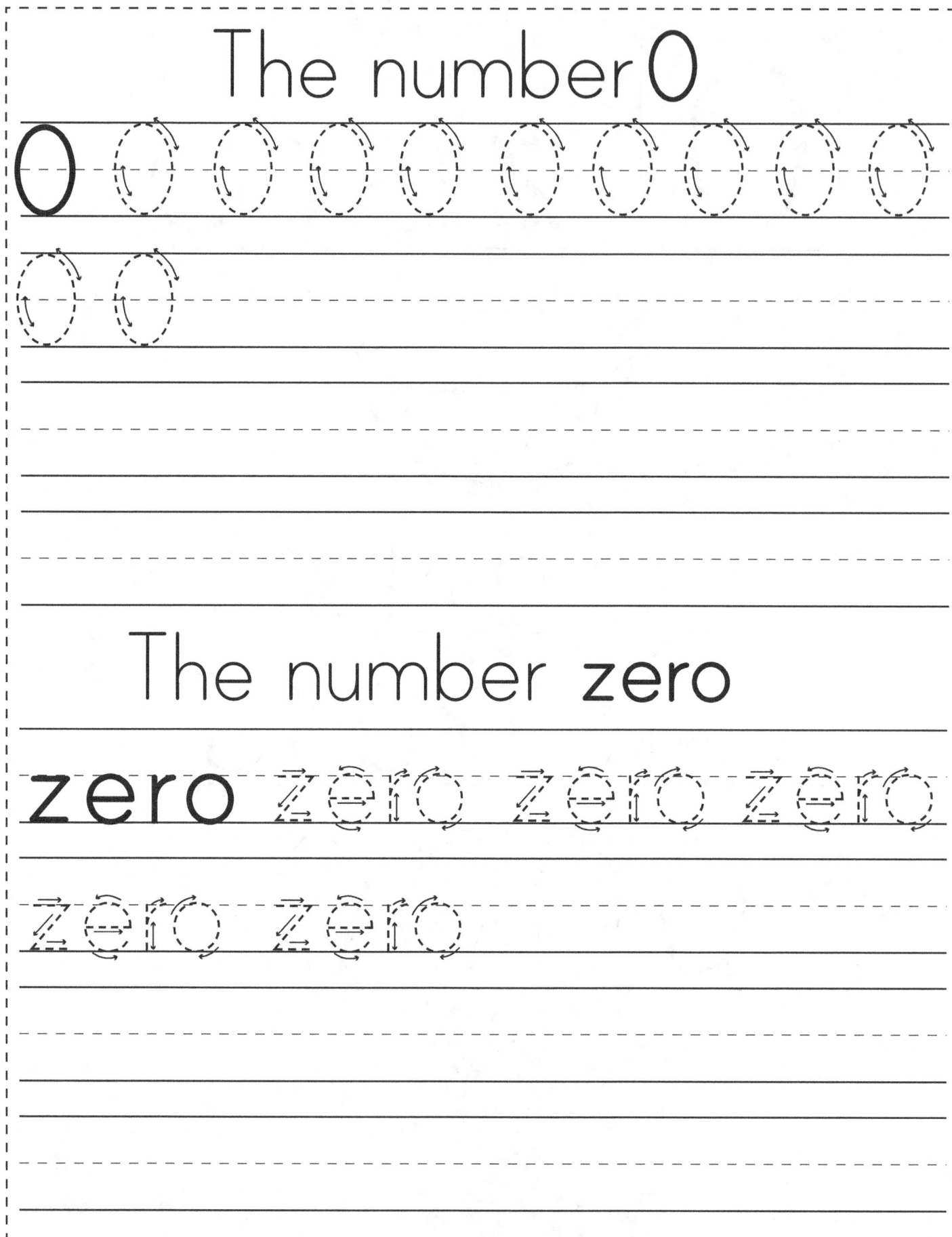

1 one

candy

one candy

I

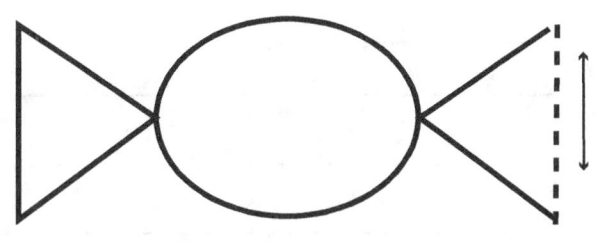

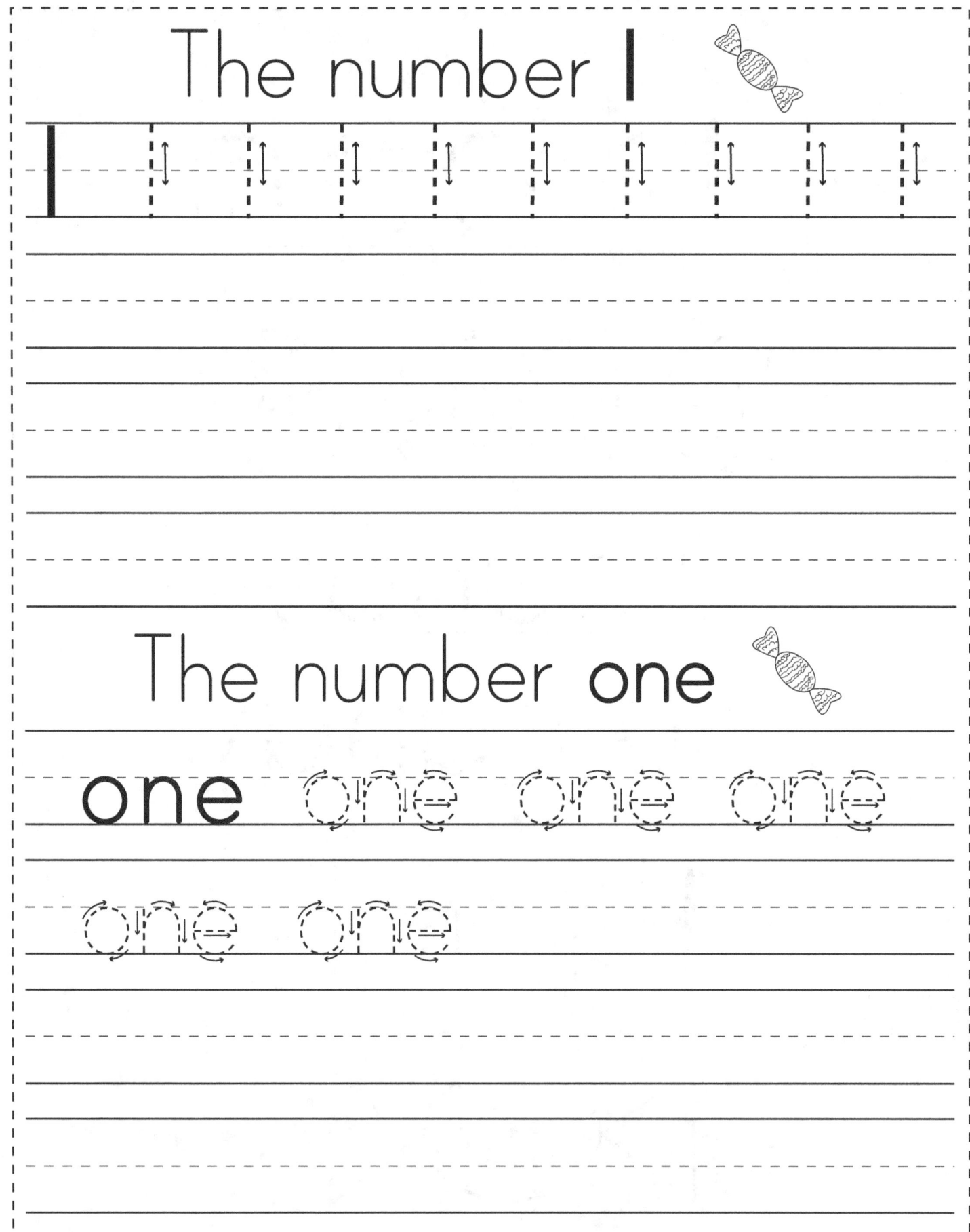

| 2 | two |

2 swans

two swans

2 2 2 2

| 3 | three

3 butterflies
three butterflies

3 3 3 3

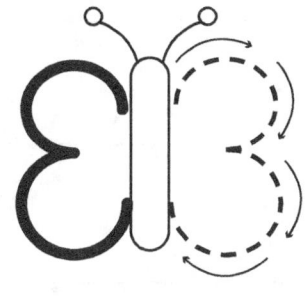

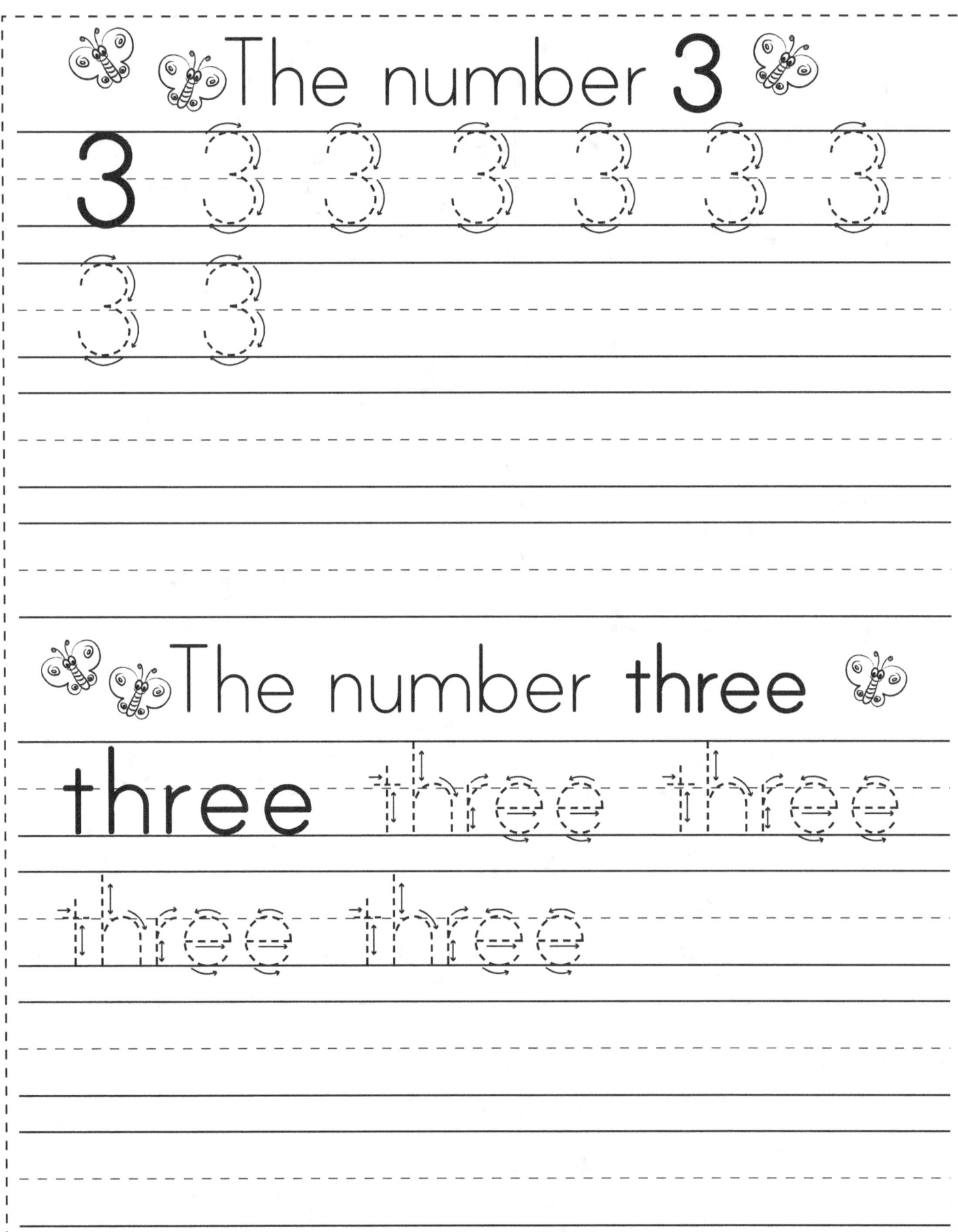

4 four

4 chairs

four chairs

4

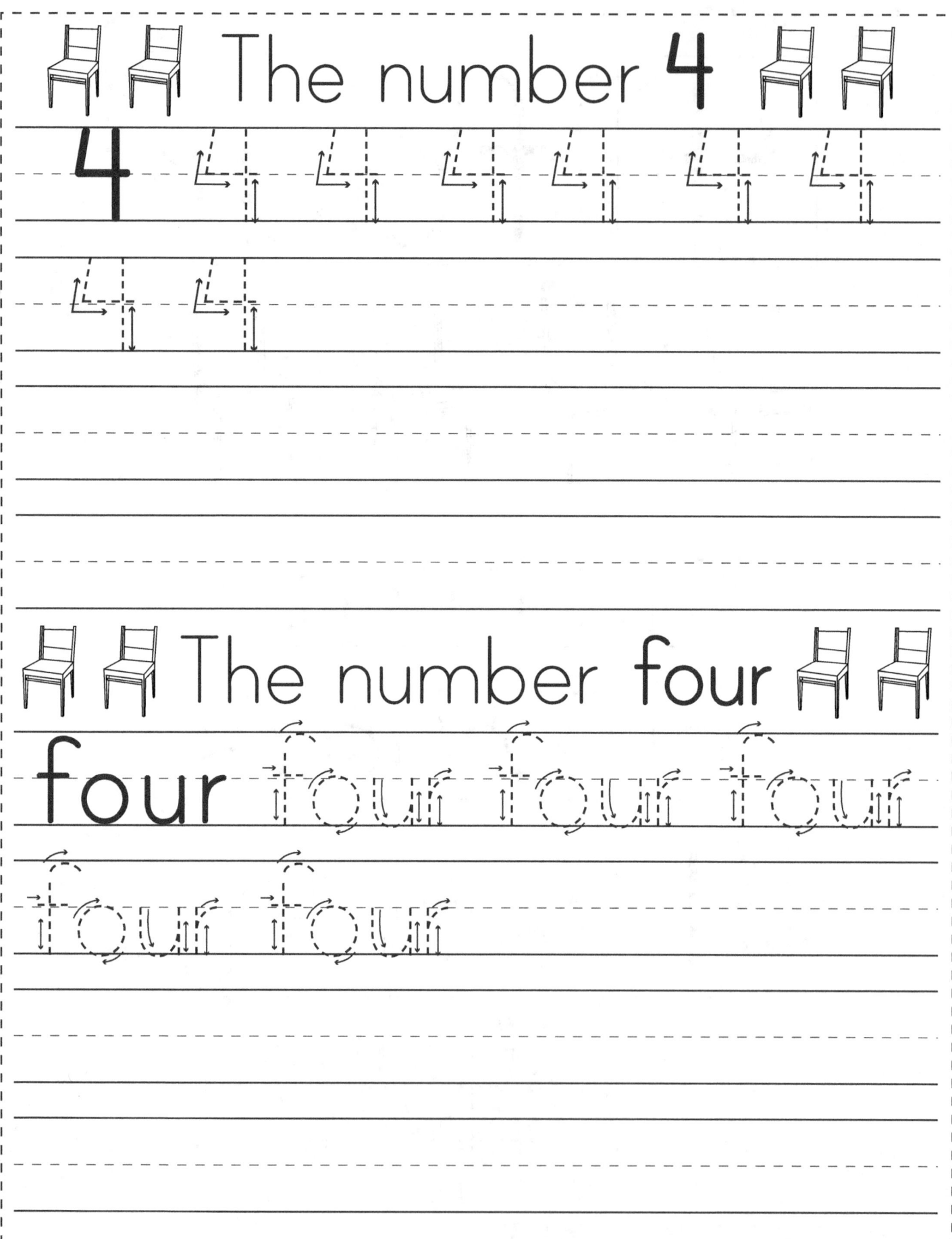

The number 4

4 4 4 4 4 4 4
4 4

The number four

four four four four
four four

5 five

5 dogs
five dogs

5 5 5 5

6 six

6 snails

six snails

6 6 6 6

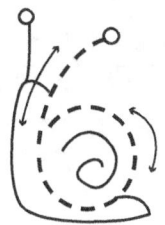

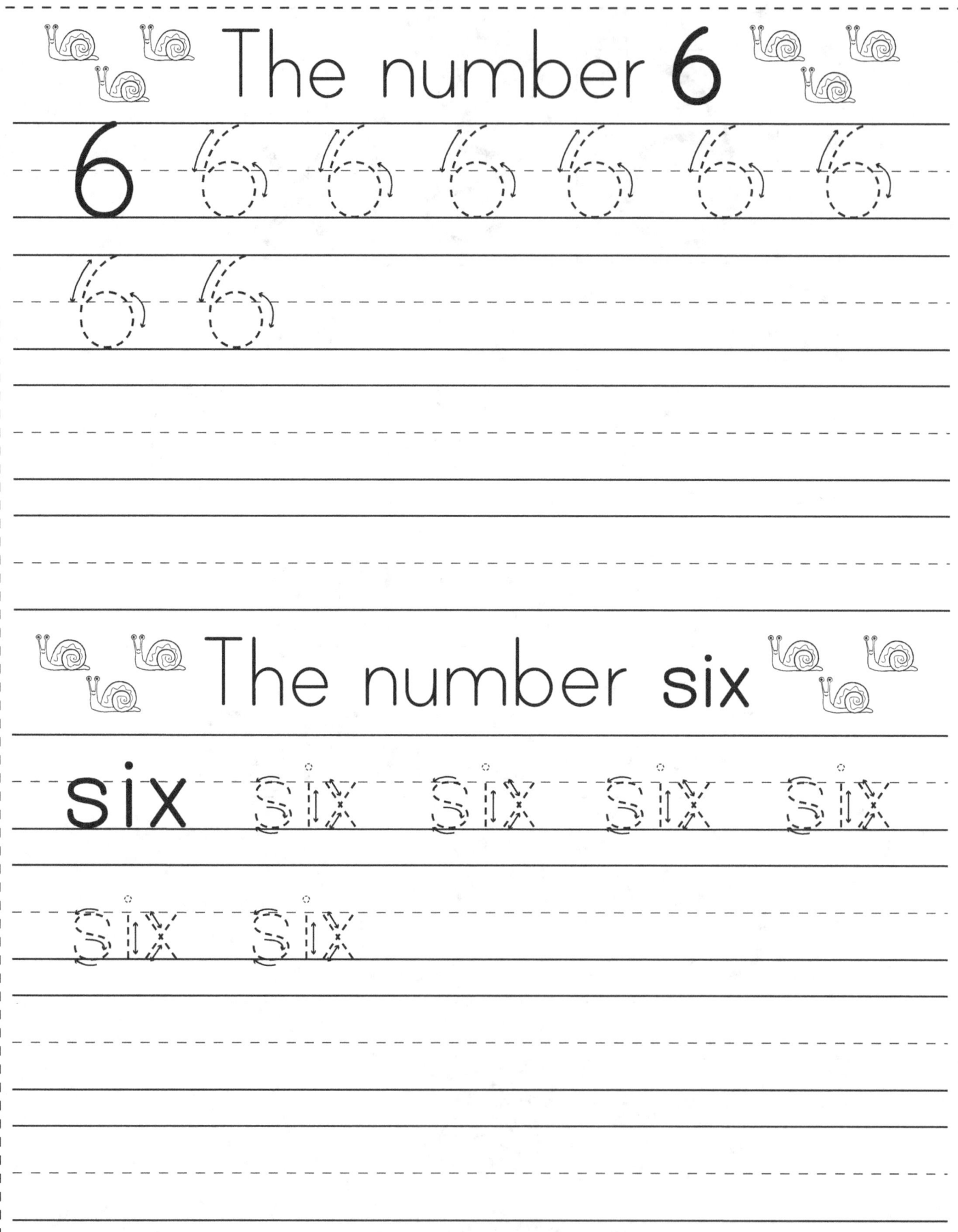

7 seven

7 cups

seven cups

7 7 7 7

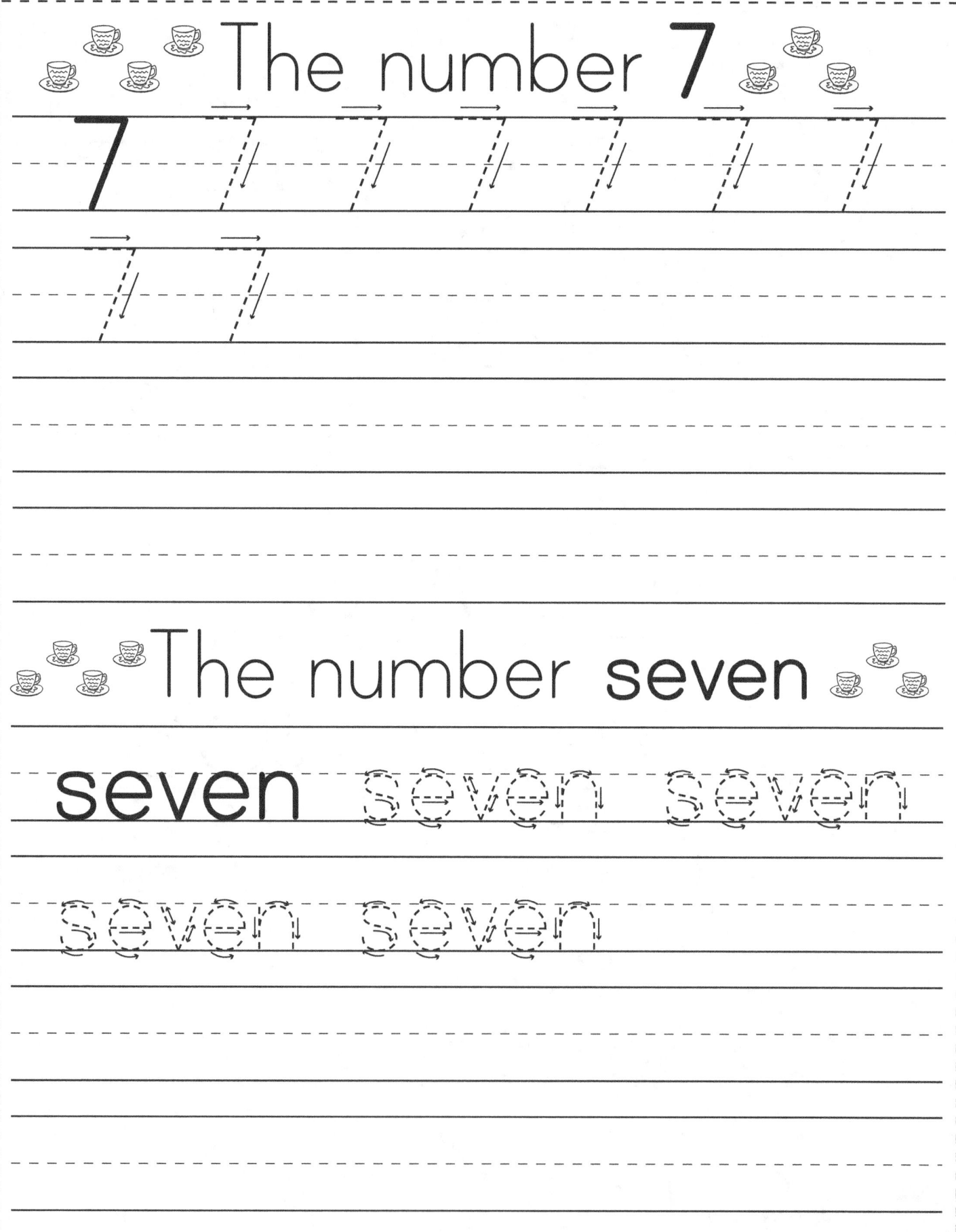

8 eight

8 bears

eight bears

8 8 8 8

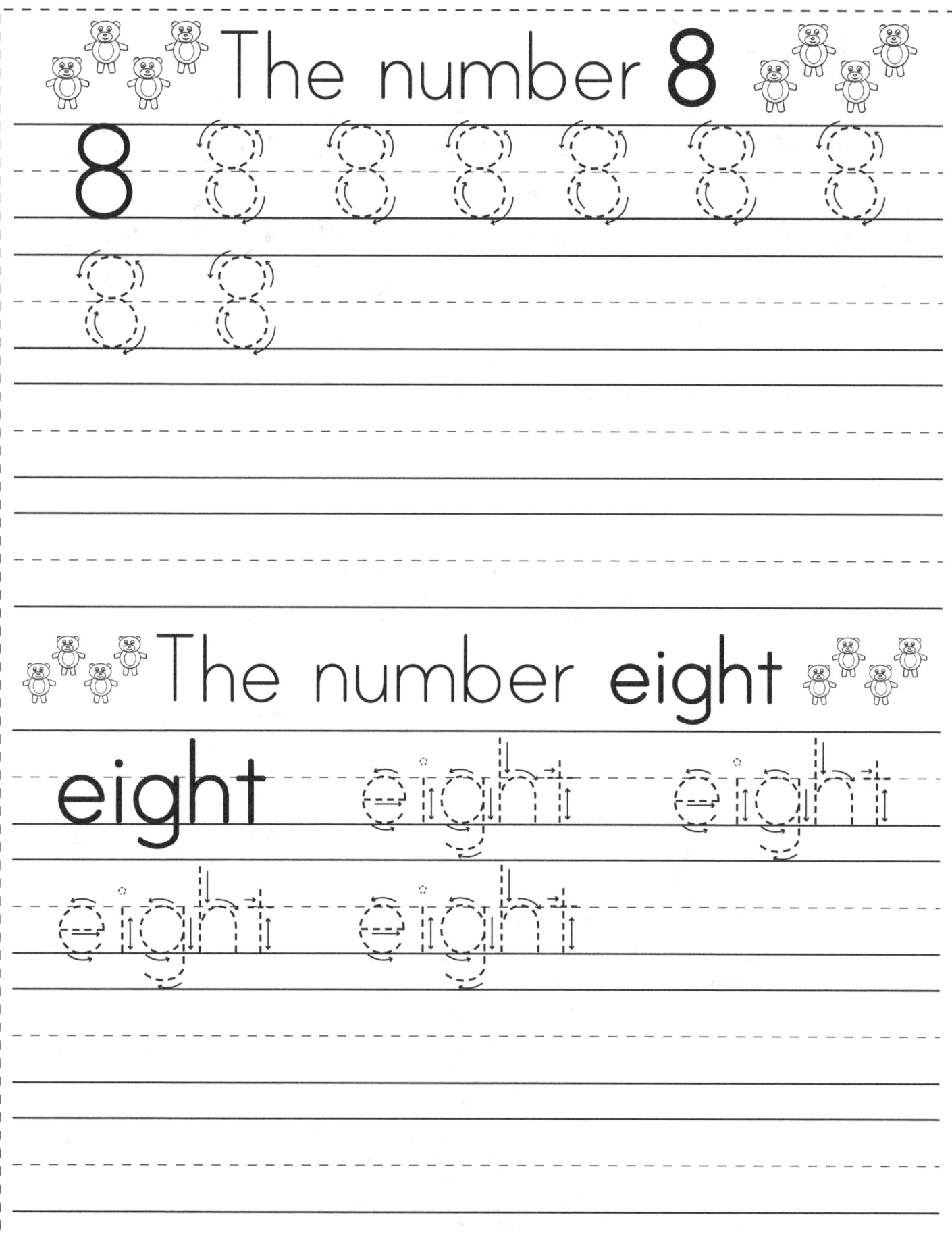

9 nine

9 owls

nine owls

9 9 9 9

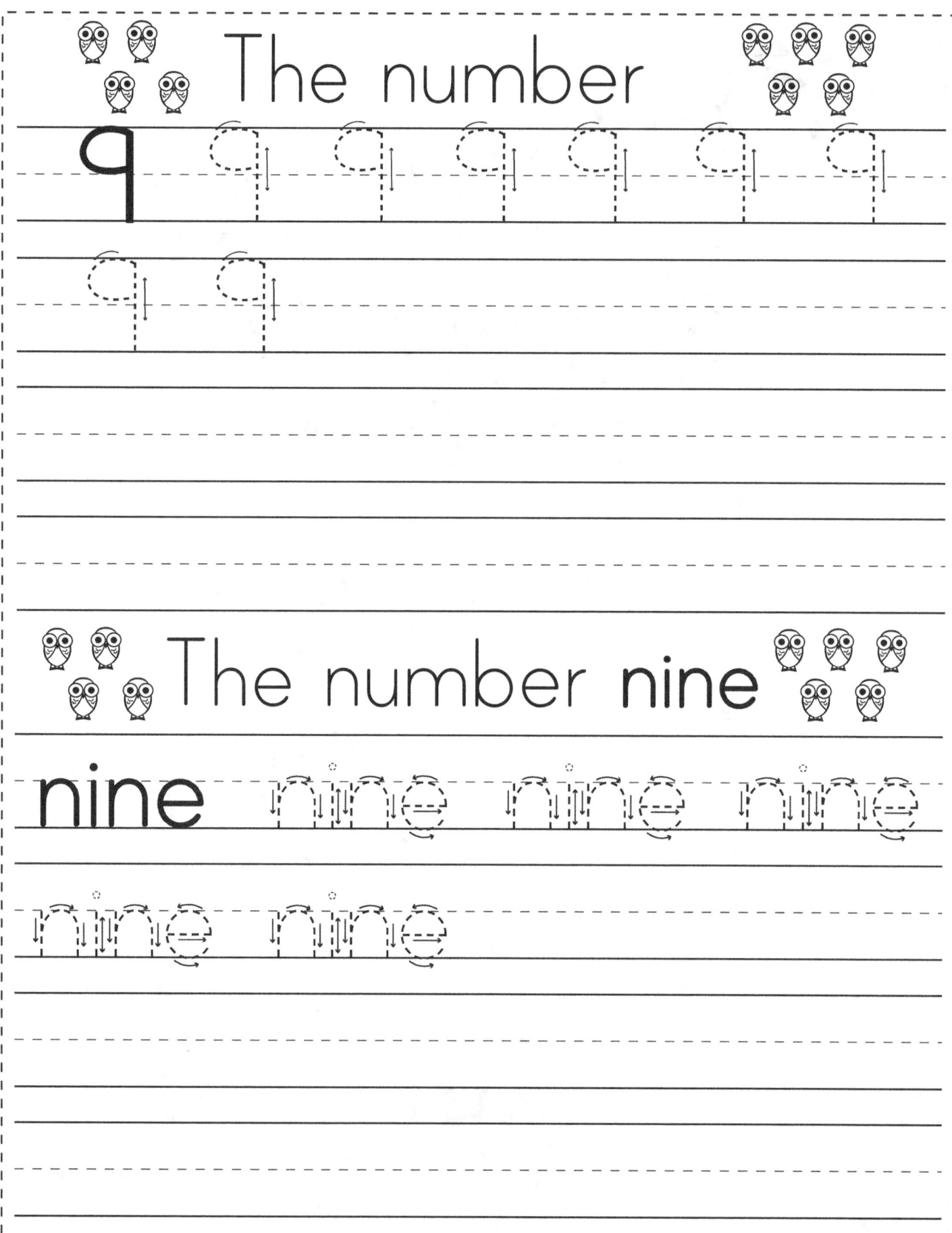

10 ten

10 cars

ten cars

10 10 10 10 10

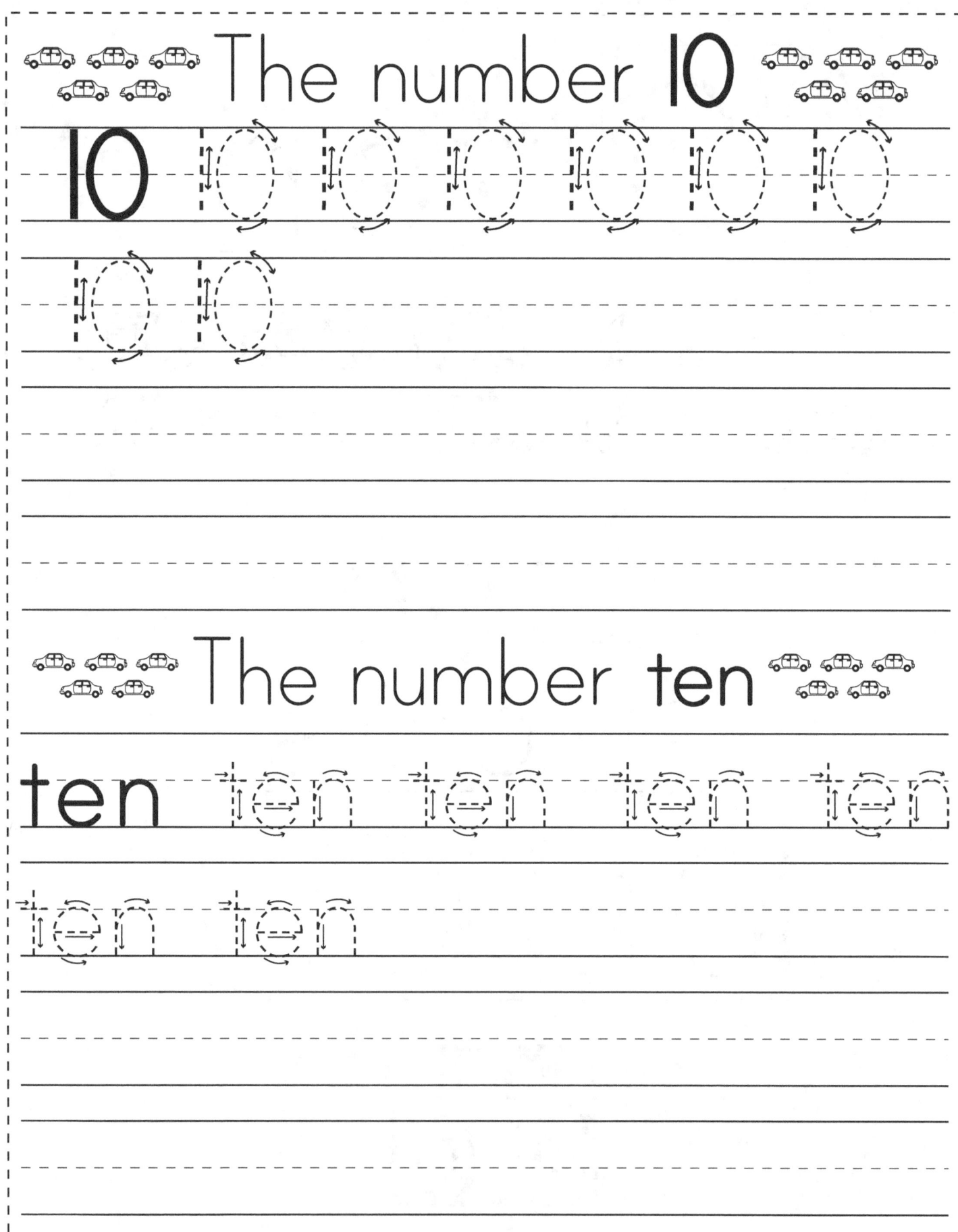

11 eleven

_11_ trees

eleven trees

11 _11_

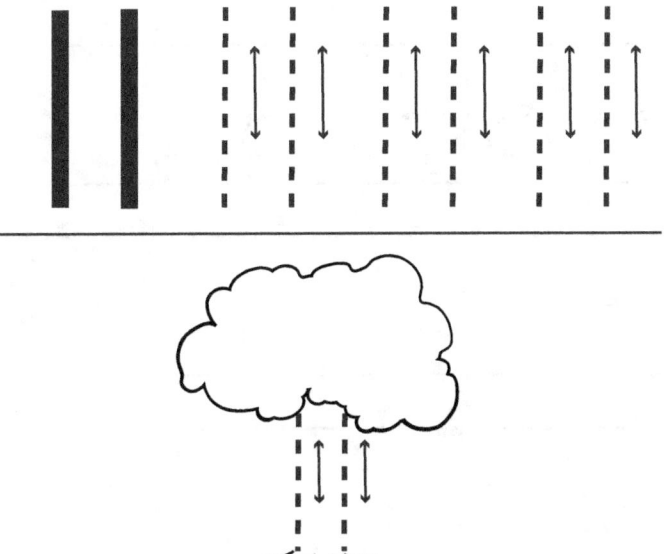

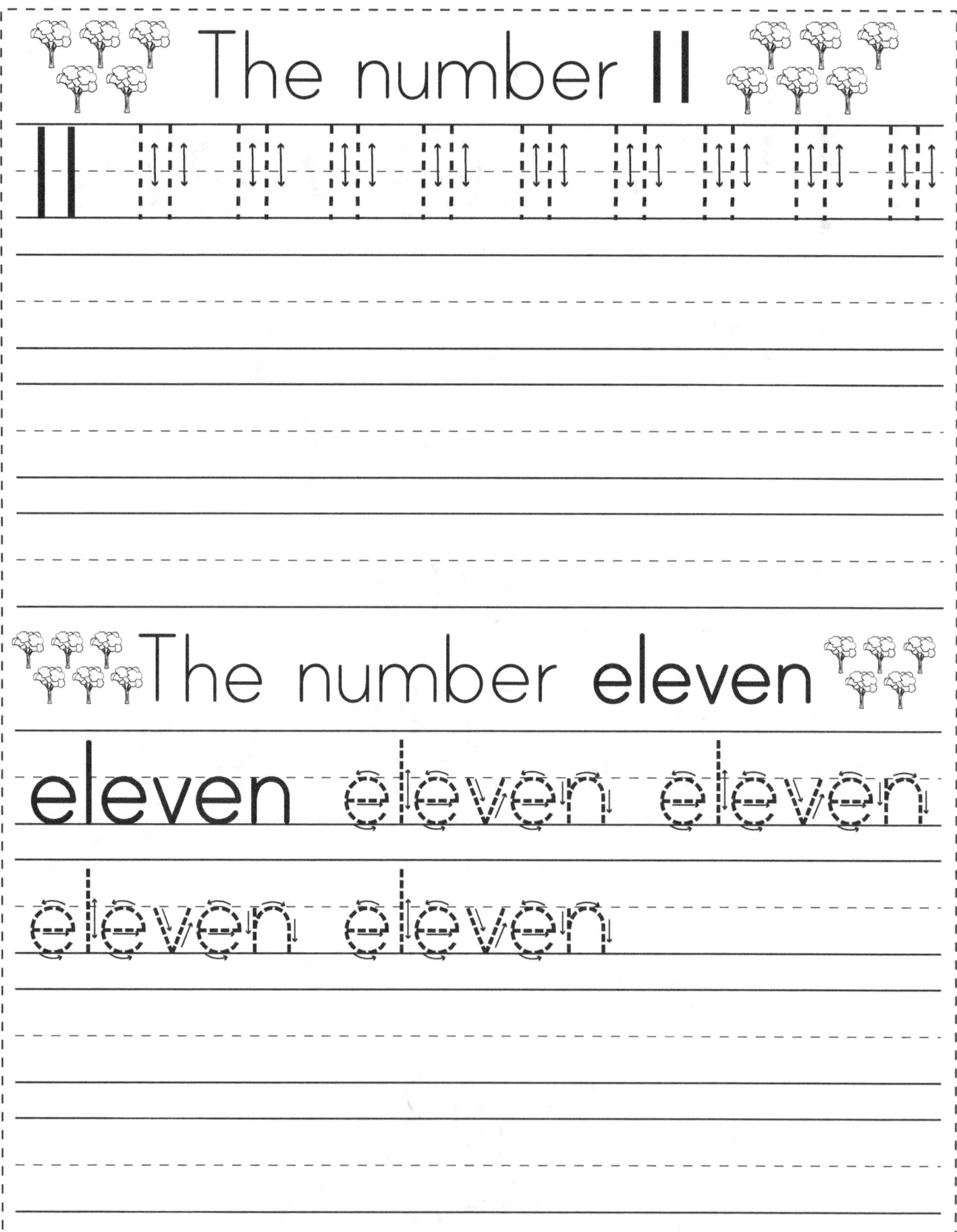

| 12 | twelve |

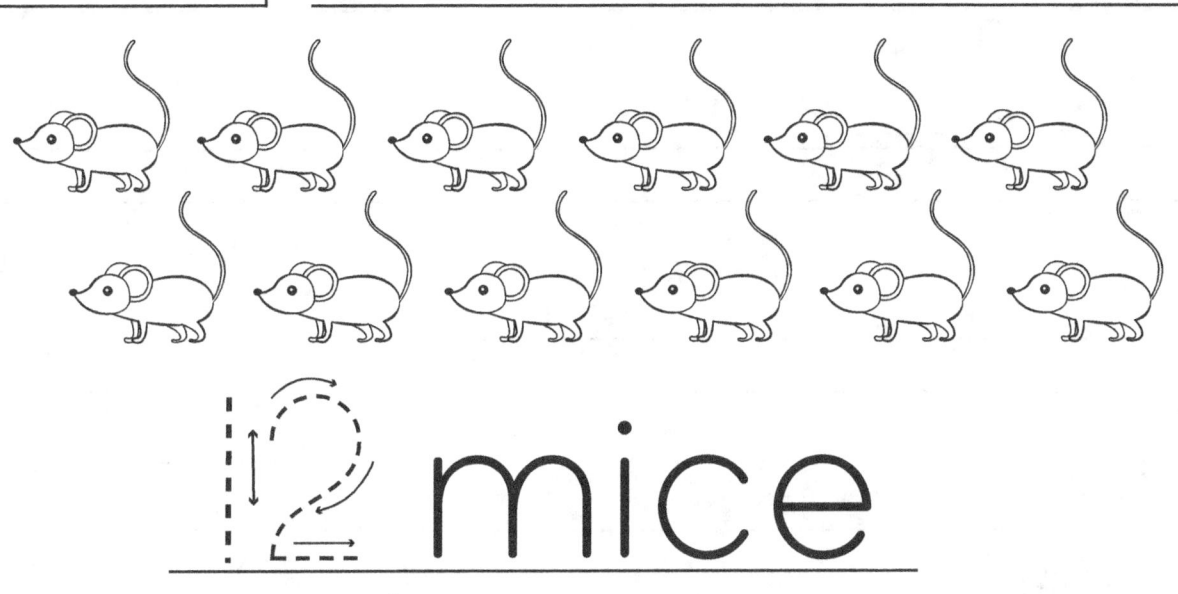

12 mice

twelve mice

12 12 12 12 12

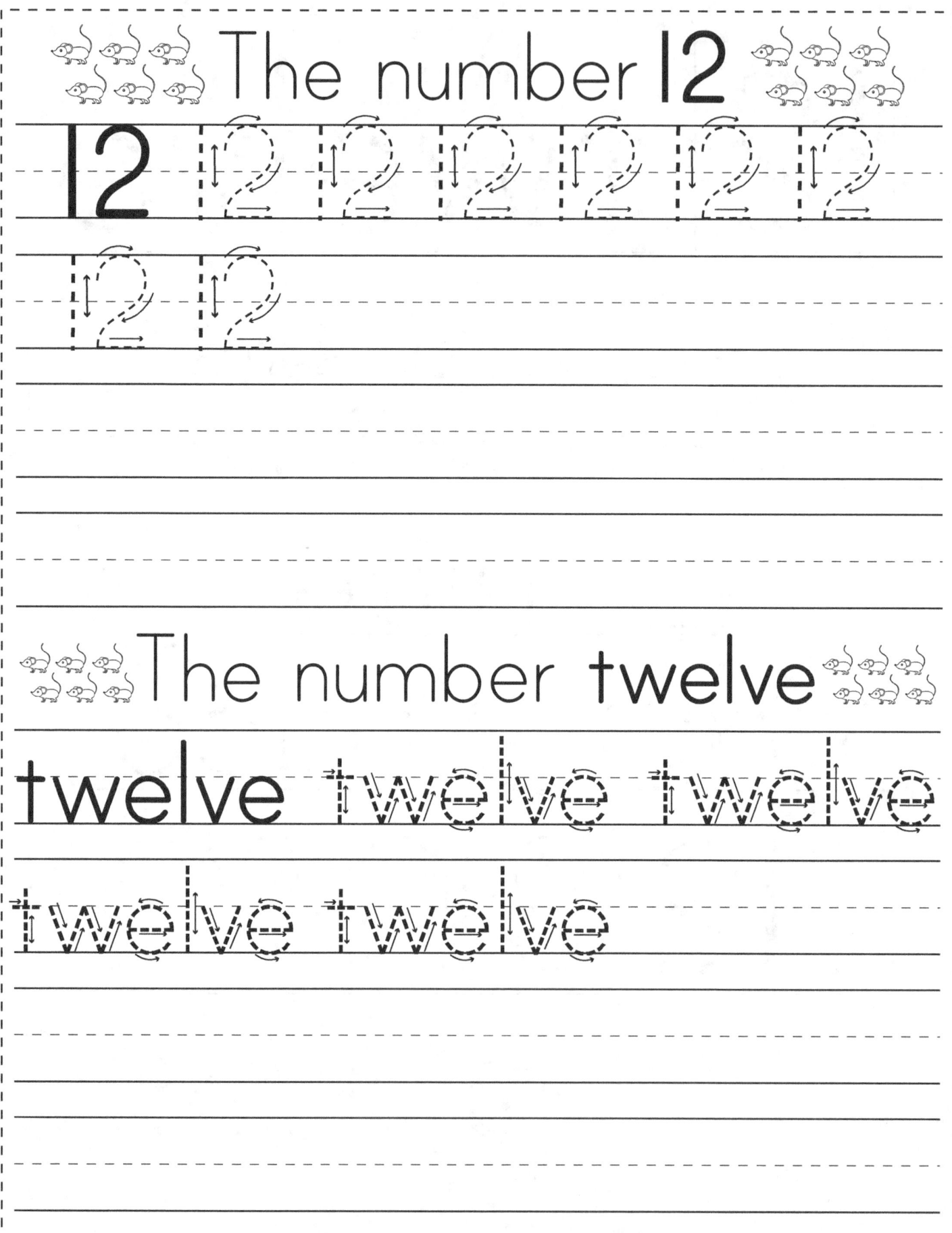

13 thirteen

13 fish

thirteen fish

13 13 13 13

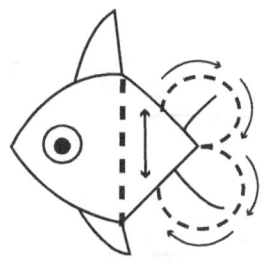

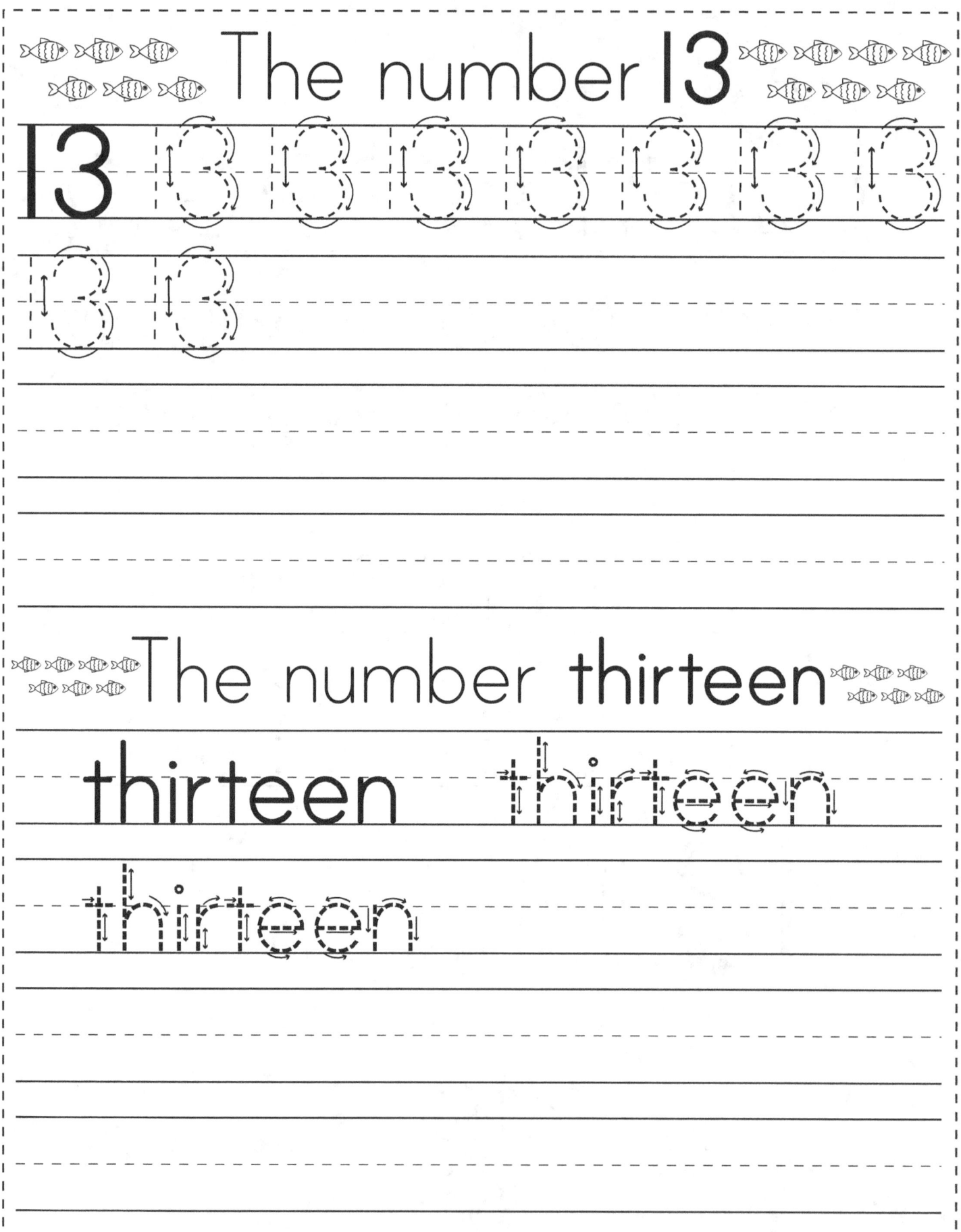

14 fourteen

14 cats

fourteen cats

14 14 14 14

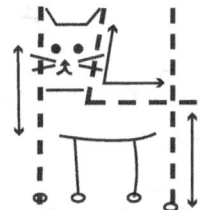

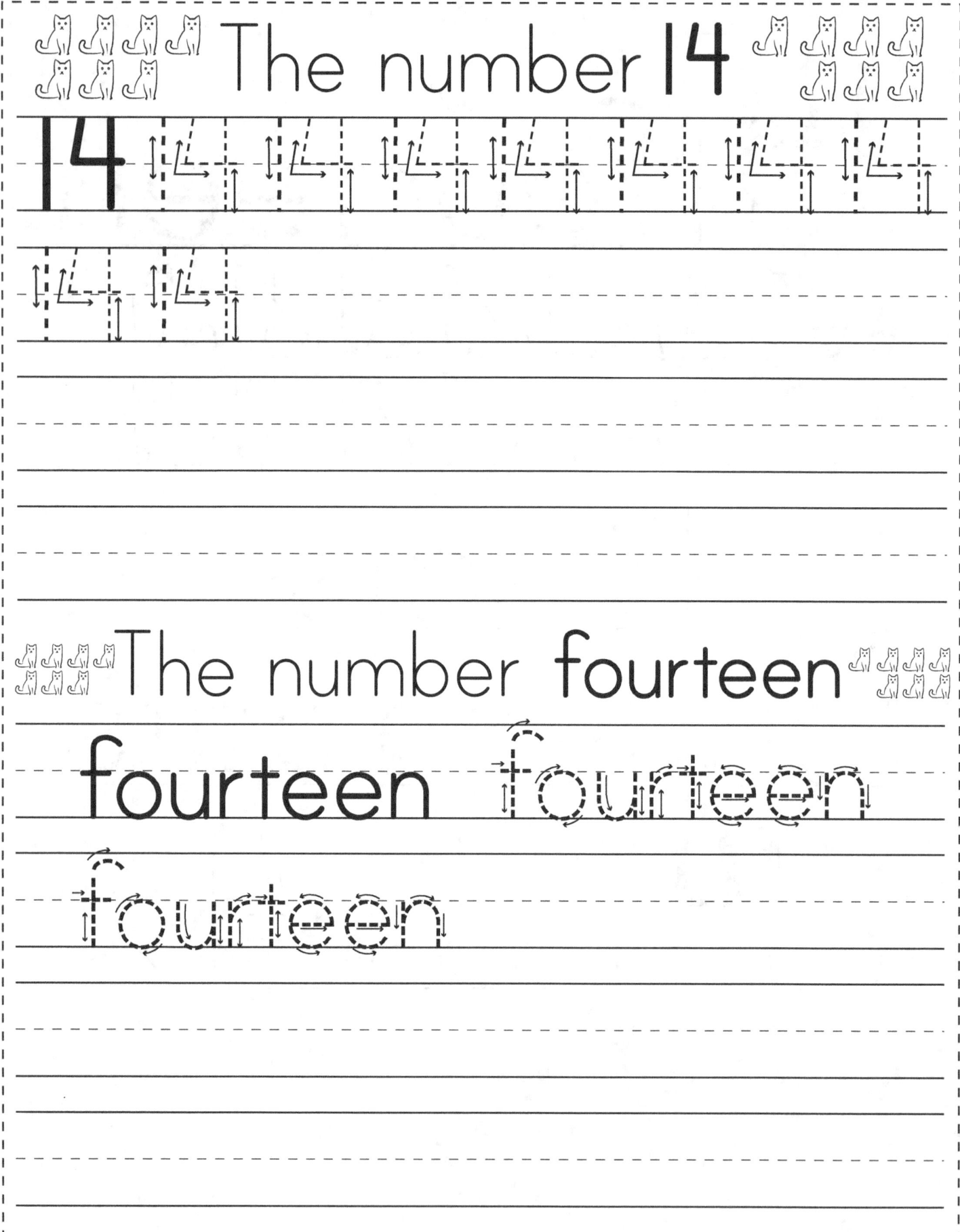

15 fifteen

15 buses

fifteen buses

15 15 15 15 15

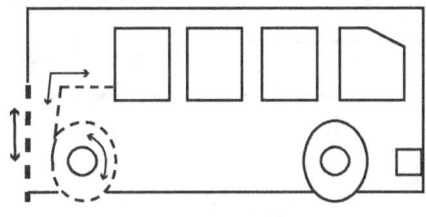

16 sixteen

16 grapes

sixteen grapes

16 16 16 16

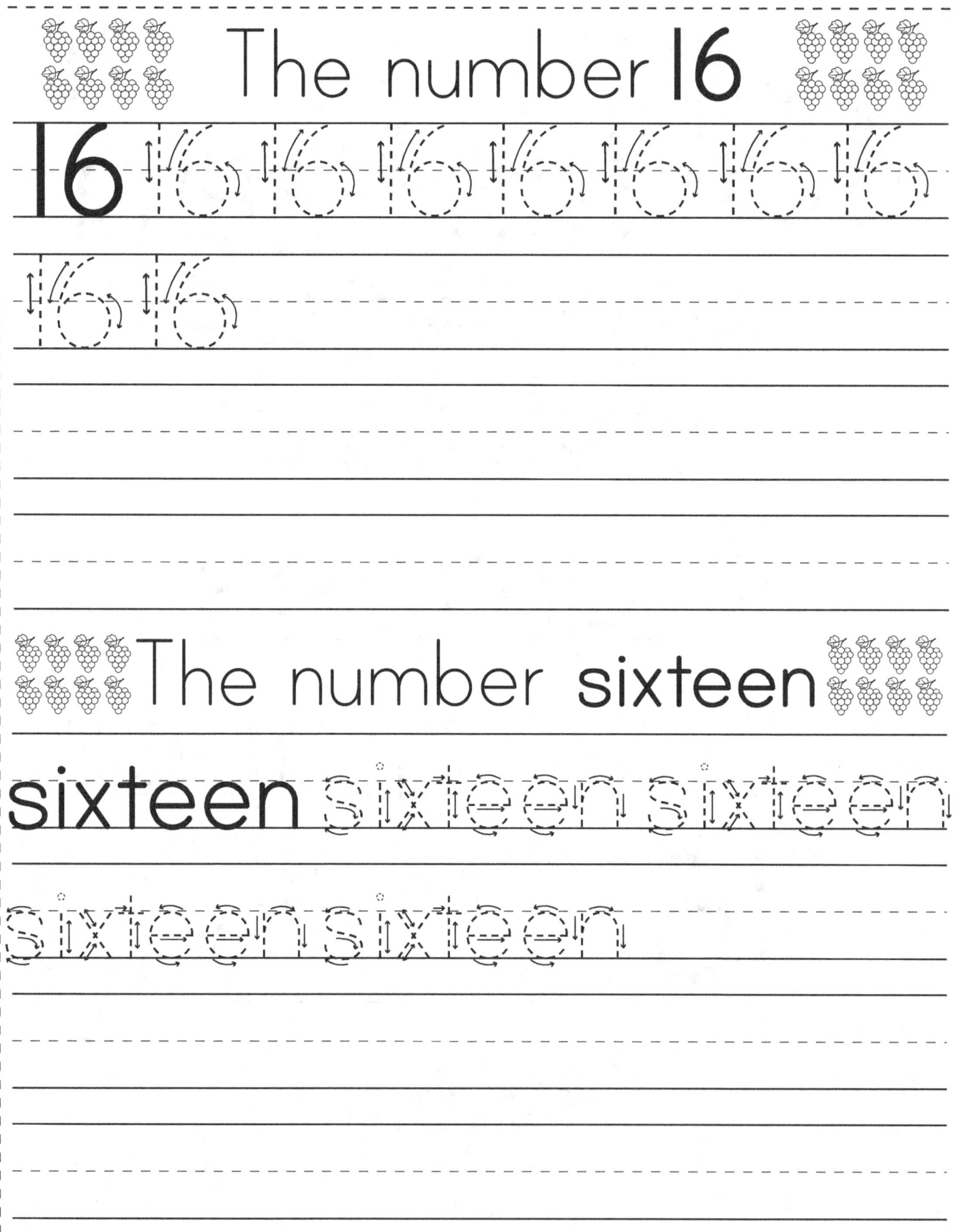

17 seventeen

17 leaves

seventeen leaves

17 17 17 17 17

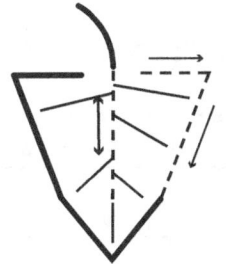

| 18 | eighteen

18 bats

eighteen bats

18 18 18 18

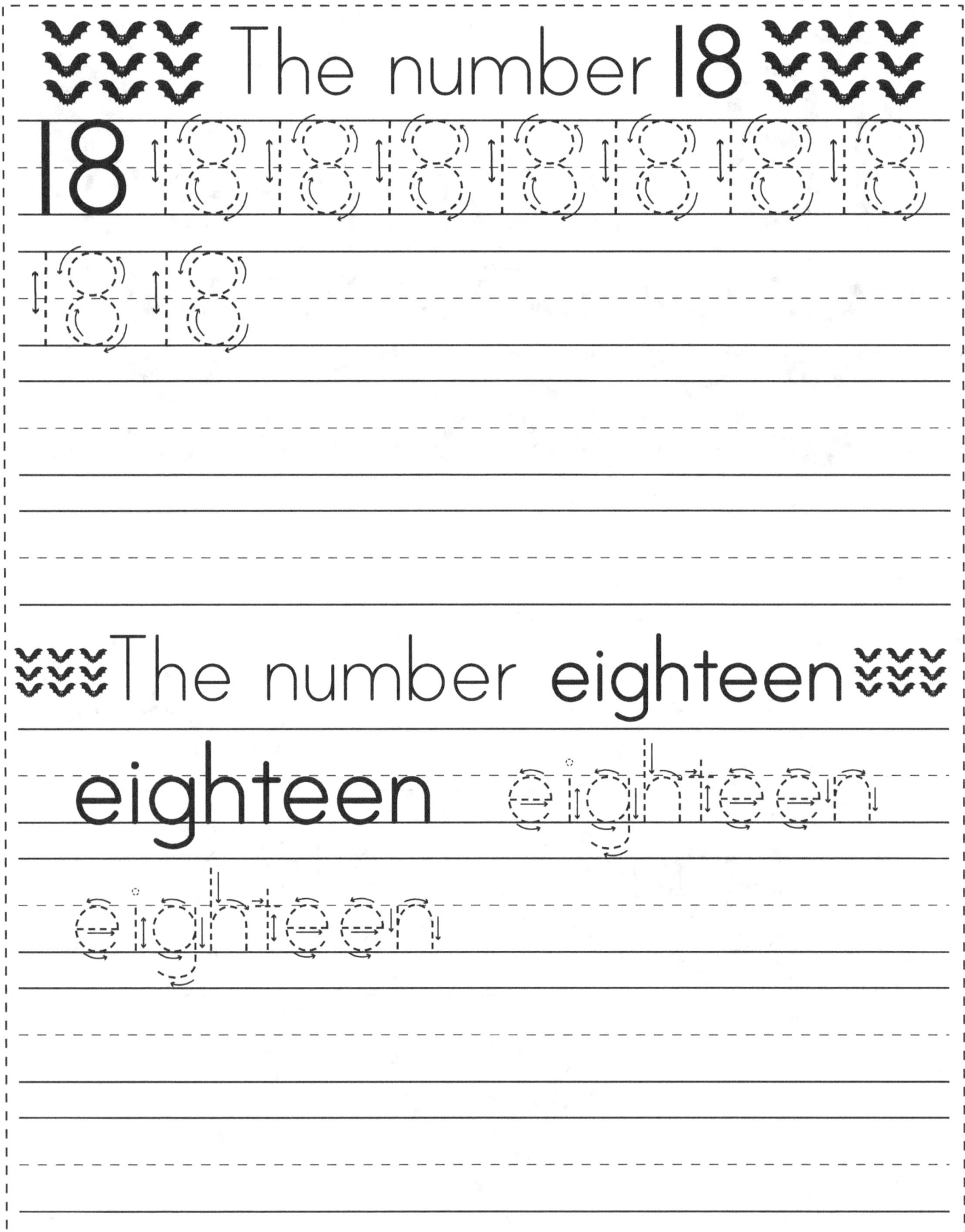

19 nineteen

19 radios

nineteen radios

19 19 19 19 19

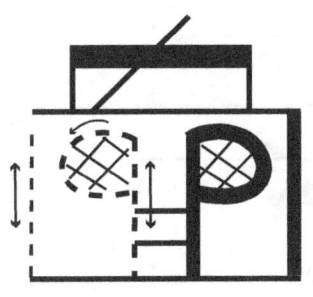

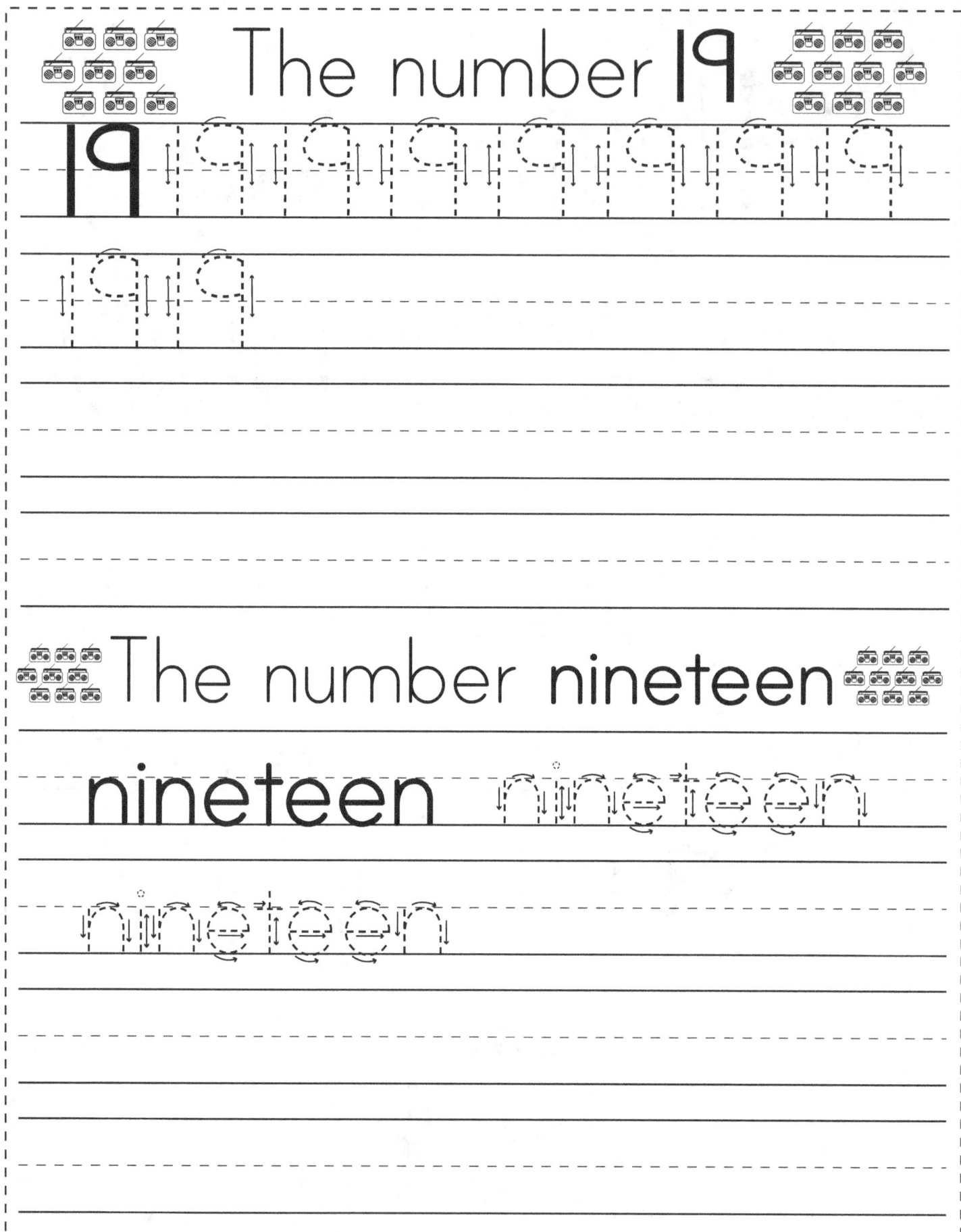

20 twenty

20 bikes

twenty bikes

20 20 20 20 20

www.ingramcontent.com/pod-product-compliance
Lightning Source LLC
Chambersburg PA
CBHW080911220526
45466CB00011BA/3543